Jarrold Bird Series
with text by **Reg Jones**

Birds in our Gardens

Jarrold Colour
Publications, Norwich

Introduction

The greater part of our environment reflects the influence of man to some degree. Fields enclosed by hedgerows, coppiced broad-leaved woodlands, and forests of regimented pines are examples of the way in which the land has been modified by his actions. However, nowhere is his capacity to effect change more evident than in our towns and cities. Some are bleak and bare, but fortunately it has been almost traditional, especially in Britain, to associate open spaces with buildings. There are timbered squares and parks, and, on a more personal basis, many dwellings have gardens. Their provision may have been an attempt, perhaps unconscious, to link the inhabitants with their distant rural origins.

It is a simple matter of observation that gardens are populated by birds and the question may be asked, 'Where have they come from?' Undoubtedly the majority are woodland in origin and include those species which have been sufficiently adaptable to exploit the possibilities for feeding and breeding within a garden. In addition, temperamentally they have been able to tolerate the close proximity of man. Blackbirds and robins have been very successful in making the transition whereas the woodland warblers have not enjoyed the same degree of success.

Of course, the number and kind of birds seen varies considerably. The pocket-handkerchief-sized plots, sometimes reduced to paved yards, which are typical of the tightly packed terraced houses built towards the end of last century may attract little other than house sparrows, starlings, and feral pigeons. There may be food in the form of household scraps but little cover and nesting-sites are confined to spaces offered by deteriorating brickwork and poorly maintained roofs. Large detached houses built in the same period for the professional and managerial classes were often surrounded by large gardens, luxuriantly endowed with shrubs and trees, and where these still exist in central areas they function as oases for a greater variety of birds.

During the present century, and particularly since the end of the First World War, there has been a steady spread of housing development from centres of population into the encircling countryside creating suburbs. Most of these newer dwellings have gardens and the abundance of the bird population is determined in large measure by the extent to which there are well-established bushy boundaries, flourishing shrubberies and sizeable trees. Additionally the close proximity of tree-lined avenues, open farmland, scrubland, golf-courses, or parks indirectly influences the diversity of the birds which appear from time to time in the actual gardens.

Greenfinch feeding on peanut

House Sparrows, which are common in the smallest of urban gardens, are not woodland birds which have changed their way of life to exist in close proximity to man. True sparrows, of which the house sparrow is one example, are generally regarded as being related to weaver birds. They are seed-eating and live especially on the seeds of grasses and cereals. It is suggested that the ancestors of sparrows originated in open country with scattered trees or bushes in which domed, weaver-like nests of dead grass, lined with feathers, and with side entrances, could be built. This ancestral group probably evolved in tropical Africa and reached Europe and Asia via the Valley of the Nile.

The house sparrow is a species which has formed a close connection with

Left: *Cock House Sparrow* (×1). **Below:** *Hen House Sparrow* (×1).

man and has extended its range as he has settled down to a sedentary, agriculturally based life, utilising his cereal crops, and later his household scraps, for food. Sometimes untidy nests are placed in bushes, but more frequently birds take over holes and crevices in buildings and these are lined to form enclosed shelters. House sparrows are now found throughout Britain, having become established in the northernmost areas and in the Western Isles only within the last hundred years.

A cock house sparrow is distinguished by its grey crown which contrasts with the darkly streaked reddish-brown feathers of the upper parts. The throat and breast are grey except for a black bib. The hen is not so warmly coloured and she lacks the grey crown and the dark patch on the throat.

Below: *Hedgesparrow or Dunnock* ($\times \frac{2}{3}$). **Right:** *Robin* ($\times \frac{2}{3}$).

At first sight the Hedgesparrow or Dunnock may be thought to resemble a house sparrow, but dunnocks belong to a different group and are not true sparrows. The sexes are alike and although their backs are warm brown in colour, their heads are grey and their flanks noticeably streaked. The bill is much finer than that of a house sparrow, indicating a preference for an insect diet.

Hedgesparrows are usually found in open woodland, scrub, or hedgerows where there is plenty of low cover created by brambles or nettles among which simple cup-shaped nests of thin twigs and moss are hidden. The eggs are spotless pale blue. The species is common in gardens; dense hedges such as those of privet providing the necessary concealment.

While house sparrows are social birds and occur in parties, the hedgesparrow is curiously detached and solitary. It does not attempt to join tits on bags of peanuts and rarely appears on bird-tables, preferring to hop inconspicuously below collecting trifles which have fallen to the ground. It is constantly on the move and never becomes confiding. Its call, which is usually uttered while hidden, is a squeaky 'tseep' and the song is a simple trill, much more subdued than that of a wren.

No bird is more familiar to the suburban gardener than the Robin. As soon as soil is turned and its surface broken, a robin will appear as if by appointment and perch on some vantage point as close as possible to the place where work is proceeding so as not to miss the opportunity of securing some freshly exposed and temporarily defenceless grub. Most gardeners have experienced this companionable relationship.

Robins never achieve the same numbers as house sparrows, starlings, and blackbirds and, in fact, being very conscious of territory, it is not usual in winter for more than one bird to be evident in any one place. They are not often present in the tiny gardens of built-up central areas but occur in larger plots which offer a varied environment. Small trees are used as song-posts and well-grown shrubs for cover. Nests are hidden among creepers or within niches or, not infrequently, in discarded receptacles in untidy corners. Some birds venture indoors and, on occasion, construct quite massive nests in sheds and garages.

A robin's song is a thin liquid warble and somewhat melancholy. It characterises autumn and also, in a bolder version, announces the arrival of spring.

In gardens of any size, Blackbirds are often the commonest birds excepting house sparrows. As in woodland, they take much of their food from the ground and they are particularly conspicuous when foraging on lawns. Running along, they pause with heads cocked, attempting to locate with certainty hidden worms or grubs before attempting to secure them with darting stabs. They feed largely on animal material supplementing their diets in late summer and autumn with berries such as those of cotoneaster. They are familiar visitors to bird-tables.

Blackbirds are woodland in origin but, at the same time, they are extremely adaptable and this has allowed them to move into other habitats, including gardens, during the last century. Their adaptability is very apparent when

breeding. Nests are usually built in shrubs or trees but some are placed on suitable ledges, or in cavities, within garden-sheds, workshops, or garages. No site offering any kind of security is neglected.

The handsome cock bird, normally completely black except for a yellow bill and eye rim, is one of the finest of garden songsters. In spring, at dawn and in the evening, his mellow fluting is well known, the phrases of the song being longer and more fluent than those of thrushes. The hen is not so distinctive. She is dullish brown and slightly speckled on the breast. Both birds utter a soft 'chook-chook' to register concern and when startled or alarmed produce an explosive stream of higher pitched notes as they dash for cover.

Left: *Cock Blackbird* ($\times\frac{2}{5}$). **Above:** *Hen Blackbird* ($\times\frac{2}{5}$).

The Song Thrush is not normally present in the smaller and more centrally placed gardens nor does it achieve the same numbers as the blackbird in the larger gardens of the outer suburbs. In fact it is usually outnumbered by more than five to one. This may be because the song thrush is not so assertive as the blackbird and also it restricts itself for breeding to dense hedges and bushes within which it secretes its mud-lined nests.

The sexes are alike, fawnish brown above and with lighter underparts streaked with dark brown spots. There is a small patch of buff on the flanks and also beneath the wings, the latter being visible during flight. The song is more highly pitched and bolder than that of a blackbird, and is characterised by single notes and short phrases which are sometimes repeated several times.

As with blackbirds, song thrushes collect most of their food from the ground. They unearth many worms and ferret out snails from their day-time

Left: *Song Thrush* ($\times\frac{2}{5}$).
Right: *Mistle Thrush* ($\times\frac{1}{3}$).

shelters, breaking the shells by battering on stones before swallowing the softer nutritious parts.

The Mistle Thrush is a casual visitor to suburban gardens. Appearing on a lawn it can be seen to be a larger bird, to carry itself more upright, and to be more forceful than a song thrush. In addition, it is greyer and lacks the traces of buff apparent on the smaller species. On the wing, apart from the impression of greyness, white tips are apparent on the outer tail feathers.

Mistle thrushes usually build their nests in the forks of trees at a good height from the ground. Few gardens have trees which are tall enough and sufficiently lofty specimens are more likely to be found in adjoining parks or plantations. Mistle thrushes nest early in the year and at this season cock birds sing from elevated perches, often, in wild weather, repeating three or four syllabic phrases with almost defiant intensity. There is good reason for referring to them as Storm-cocks.

Apart from house sparrows and blackbirds, Starlings are usually the commonest birds occurring in the majority of gardens. They perch on TV aerials and chimney-pots pouring out a medley of sounds which include penetrating whistles, harsh nasal 'tcheers', together with an extensive repertoire of different calls, many of which are imitations of other birds. A handful of breadcrumbs brings them to the ground where they are boisterous and aggressive. At other times they busy themselves on lawns, vigorously probing for grubs, and leave the surface pitted with shallow craters.

Starlings breed in cavities. In woodland they take over holes which have been produced by decay or by the activities of woodpeckers, whereas in towns most birds nest beneath roof tiles.

Both sexes are black, the plumage possessing an iridescence which shows as purple and green hues. In spring the feathers on the back and

Left: *Starling* (×⅔). **Right:** *Wren* (×¾).

wings are pale tipped but by autumn, after completion of the summer moult, they are speckled all over. The majority of the light flecks are lost as the winter progresses. The dark colour of the bill also changes to yellow before breeding starts.

Wrens are very adaptable. Probably they are most at home in damp broad-leaved woodlands with an abundance of cover as is provided by brambles and briers rampaging over decaying stumps and fallen branches. Here they build their dome-shaped nests, often using the dead fronds of ferns to form the outer shells, placing them within a framework of rotten timber or suspending them between thorny strands. However, wrens may be encountered in a variety of other situations where there is sufficient concealment. It may be provided by parkland shrubberies or, at the other extreme, by the spaces between the boulders littering the sides of mountains.

In winter, wrens are occasional visitors to many suburban gardens. Russet-brown in colour, diminutive yet plump, they are constantly on the move hunting over the ground for insects which form the main part of their food. Their tails are usually held in an upright cocked position. Normally they only stay to breed where there are dense, and often neglected, areas of greenery. Sometimes they may find refuge in an outbuilding, constructing their nests within some discarded article such as a hanging rotten sack.

Left: *Blue Tit* (×1). **Below:** *Great Tit* (×⅔).

Blue Tits and Great Tits are present at all seasons, being constant visitors in hard weather to bags of peanuts or sectioned coconuts, while in summer both species may breed. They are woodland birds and normally nest in trees using holes or gaping clefts. Many gardens do not have timber of sufficient age to provide such sites and nest-boxes fill an obvious need. These appear to be utilised more readily by the blue tits, a pair of great tits often preferring a chamber which is reached down a vertical chimney. Here, although hidden from view, a great tit shows its resentment to disturbance by producing a scolding hiss which allows it to be identified.

Apart from size the great tit is differentiated by a shining black cap and white cheeks together with a broad black band on the yellow breast. Its most characteristic song is a seesawing 'teecher-teecher-teecher'. Blue tits may sometimes have a slight dark stripe on the breast but it never assumes the bold dimensions of that of a great tit and the smaller bird is especially noted for the blueness of the crown, flight feathers, and tail. Its song is a simple inconspicuous trill, 'tsee-tsee-tsee-chchchurr'.

Both species take large numbers of small moth caterpillars in spring and early summer when they are feeding young. They collect them at a tremendous rate, a single pair accounting for several thousand while rearing a brood.

The Coal Tit is at home in coniferous woodland. Sometimes it visits gardens in winter and feeds with other tits on food put out to attract them. Usually it is in a minority.

Coal tits are the smallest of British nesting tits and although black headed like great tits are easily recognised by the elongated white mark on the nape of the neck. The upper parts are drab olive-grey and below they are a dirty white with a touch of rufous on the flanks. Alongside the more brightly coloured blue and great tits they can look somewhat scruffy.

Coal tits also nest in holes which are very often at, or near to, ground-level. They rarely breed in gardens but can be present in adjacent spinneys especially if there are conifers. Their presence in April can be deduced from the song which is a more piping version of a great tit's.

The Nuthatch is typical of open woods and parkland with mature trees. It is commonest south of a line drawn from the Mersey to the Wash and distinctly local in more northern counties. Not surprisingly it is only present at all seasons in spacious gardens which are well timbered. In less suitable areas, nuthatches are casual visitors to bird-tables in winter.

A nuthatch is not easily confused with any other bird. Short bodied and chubby, it shows great agility on the boles and branches of trees, moving jerkily on its sturdy legs. It is unique in being capable of descending head first down a trunk. In spring its song sings out like a boy's whistle, 'pee-pee-

Left: *Coal Tit* ($\times\frac{3}{4}$). **Above:** *Nuthatch* ($\times\frac{1}{2}$).

pee', and its call is a sharp 'twit-twit' which can be mistaken for that of a woodpecker.

As the name suggests much of this bird's food consists of nuts, hazel nuts in autumn, which are wedged into crevices and chipped open by vigorous blows from the powerful pointed bill. When feeding chicks, many insects are taken.

Although nuthatches resemble tits by nesting in holes, they are peculiar in that the entrances must always be of a standard size and shape. The necessary adjustments are made by shaping wet mud to the appropriate dimensions.

Greenfinch ($\times\frac{2}{3}$).

Of the finches, the Greenfinch is the one most frequently seen in gardens. It is especially partial to peanuts and when mounted on strings or bags of them it is usually capable of repulsing any competitors. In the countryside greenfinches are found along the fringes of woodlands or well-developed hedgerows. Near to dwellings they favour gardens with robust shrubberies, particularly of evergreens, and tall boundary hedges within which they can build their bulky nests just out of reach from the ground.

A cock greenfinch is olive-green in colour, lighter on the breast and rump, and with distinct patches of yellow on the wings and on the sides of the tail. These areas of bright yellow are very evident in flight. Hens are slightly duller in coloration. Both birds have stout bills–one old name for the greenfinch was the 'green grosbeak'–which indicates that they feed in the main on seeds and fruits, preferring larger morsels as are provided by cultivated cereals, the ripe heads of groundsel and dandelion, and the fruits of trees such as elm, yew, hornbeam, and wild rose.

Greenfinches are not attractive songsters. Their commonest note is a twittering 'chi-chi-chi-chi-chit', and in spring the cock adds to this a drawn-out nasal 'swee'.

Cock Chaffinch ($\times\frac{3}{4}$).

Chaffinches are more dependent than other finches on the presence of mature trees which are used as song-posts and also as sources of food when they are rearing broods, since the leafy crowns harbour hosts of caterpillars many of which are collected by parent birds and fed to their chicks. In addition, nests are often hidden among the creepers, such as ivy and honeysuckle, encircling the trunks. As a result within the more central areas of towns and cities chaffinches are restricted to parks and to the larger and longer established gardens. In the suburbs they may be present in richly planted parts with plenty of fruit or ornamental trees.

A cock chaffinch takes over its territory in the early part of the year, calling 'pink-pink' and issuing its vigorous song of accelerating notes 'chip-chip-chip-chip-cherry-erry-erry', usually ending with a terminal flourish, 'cheeoo'. At this season he is distinguished by a slate-blue crown and bill. In autumn the crown is more buff which is the colour of the tips of the new feathers produced after moulting. By spring the tips have worn down to expose the brighter tones below. Similarly, but not by wear, the bill, which is pale in autumn, becomes blue in spring when most birds assume their handsomest dress.

A third finch, the Goldfinch, is increasing in number but as yet is not so common as the greenfinch or chaffinch. In spring, being normally resident in open country with scattered trees, goldfinches sometimes take up residence in those suburban gardens which include small orchards or a few fruit trees in which compact yet deep nests can be built, secure in the crotches of terminal branches. Perhaps this species is most evident in late summer and autumn when parties are foraging for food. A goldfinch's bill is more slender and sharp than those of other finches and it is used for extracting seeds from mature flower heads of composite plants such as thistles and dandelions. Consequently a neglected garden, badly in need of weeding, will often attract small flocks in September and October. Since the majority winter abroad, they are rarely seen in gardens at this season.

A goldfinch is less than five inches in length but extremely colourful with its red facial mask contrasting with the white cheeks. Each black wing bears a yellow bar which becomes more obvious when the bird is airborne.

While the goldfinch may be absent in the early months of the year, this is the time when Bullfinches are most likely to command attention. They are normally seen in pairs. Both birds are black capped, the handsome cock having a light grey back and warm salmon-coloured underparts, the hen being rather more drab. Late January and February mark the start of their attacks on the succulent buds of apple trees, pears, and flowering shrubs. Moving along the slender twigs, they demolish successive buds with their powerful bills before flighting to new positions. As they dip through the air they show unmistakable white rumps by which they cannot fail to be recognised.

Bullfinches build shallow and often rather slight nests in straggling hedgerows and thickets. A well-kept garden may not have the necessary cover but an overgrown and undisturbed corner suits them well.

Left: *Goldfinch* (×⅔). **Above:** *Bullfinches* (×½).

Many town squares and other public places have their flocks of Pigeons. They are feral birds, having developed as independent free-living stocks from domesticated birds which since medieval days have been housed in dovecots and lofts. Further back in time their ancestor was the rock dove. Feral pigeons are not uniform in colour, a blue-grey form with wing bars and a light rump being the closest to the ancestral dove. Some birds carry more black, others have reddish-brown markings, and a proportion show varying degrees of whiteness.

Feral pigeons forage for food in open spaces much frequented by man, including the streets and the enclosed yards and gardens of the inner suburbs. They roost and breed in buildings. A church tower often accommodates them while, in central and industrial areas, dilapidated buildings with plenty of holes and ledges, many being in semi-darkness, can house large numbers.

Woodpigeons are shy woodland birds which have penetrated urban habitats within the last hundred years. Larger and more heavily built than feral pigeons, they are characterised by a broad white band on the wing which is very noticeable in flight. A white patch on each side of the neck is readily visible when the bird is at rest.

The nest of a woodpigeon is a flimsy platform of slender twigs which is placed within the prickly crown of a hawthorn or, at a higher level, among the branches of a tree. Where city squares are planted with planes, sycamores,

Left: *Feral Pigeon* ($\times\frac{1}{5}$). **Right:** *Woodpigeon* ($\times\frac{1}{3}$).

or robinias, and thoroughfares are lined with ornamental species of Prunus, woodpigeons may breed in the late summer well hidden by the leafy canopies. Occasionally they utilise the ledges of buildings.

A woodpigeon's song is more elaborate than the simple croon of a feral pigeon and consists of five syllables usually arranged in three groups, 'coo-coo, coo, coo-coo'. It adds a pleasant rural sound to the environment. Unfortunately woodpigeons are voracious feeders and can inflict considerable damage on green crops in vegetable gardens.

Above: *Collared Dove* ($\times\frac{1}{3}$). **Right:** *Spotted Flycatcher* ($\times\frac{2}{3}$).

While both types of pigeon have been familiar to townspeople for many years, the Collared Dove, which haunts central parks and suburban gardens, is a much more recent acquisition. It was not until the mid 1950s that this species first nested in Britain, since which time it has expanded its range in an explosive manner so that twenty years later collared doves are now commonplace in suburban areas of the majority of towns and cities, including those which are heavily industrialised like Manchester and Leeds.

A collared dove is smaller and more slender than a pigeon. Its colouring is more delicate. The upper parts are light greyish buff, the head and breast being paler. Darkness is restricted to the half-collar on the back of the neck, the wing tips, and the underside of the base of the tail.

When collared doves are in residence the cock bird sings almost continuously with a monotonous 'coo-cooo-coo', the second syllable being extended and the last shortened. As he flights from one perch to another, he calls with a nasal 'kwurr'.

As with most doves, their nests are fragile platforms frequently placed in conifers.

The Spotted Flycatcher is one of the few summer visitors from overseas commonly inhabiting suburban gardens. Usually arriving in May, it will probably first be noticed sitting upright on a lower branch of a fruit tree, frequently making a swooping, swerving sally to snatch a fluttering insect before returning to its original perch. It is this habit which makes the bird easily recognisable since it lacks any particularly striking visual features, being greyish brown above and, light underneath, with dark streaks on the throat and breast.

Spotted flycatchers are not unlike robins in their choice of nesting-sites. Nests are placed in crevices in the trunks of trees, among creepers, in holes in walls, or on ledges under canopies and verandas. Open-fronted nest-boxes are often taken over. Flycatchers are not usually feeding young before the end of June when insects are plentiful.

eft: *Cock Lesser Spotted Woodpecker* ($\times\frac{1}{2}$).

Any of the three British woodpeckers may visit gardens in the outermost suburbs during the winter months. In the absence of frost the Green Woodpecker puts in an occasional appearance, especially where there are extensive lawns based on sandy soils, and explores the ground for ants.

The Lesser Spotted Woodpecker, which is not much larger than a sparrow, sometimes leaves the uppermost branches of woodland trees to visit bird-tables in hard times. The white spots on its back are arranged in regular rows so as to give the impression of bars. Cock birds are red capped while the hens lack any warm colours.

Commonest of the three is the Great Spotted Woodpecker. It makes frequent visits to gardens to sample the food put out for other species such as tits, it penetrates closer to the centres of urban areas and where there are ancient trees with some rotten wood breeding can occur. A small well-timbered park or undisturbed Victorian garden may be found to have a pair in residence. Great spotted woodpeckers resemble their smaller relatives in having red markings on the head confined to the males.

Above: *Cock Great Spotted Woodpecker* ($\times\frac{2}{5}$).

Above: *Tawny Owl* ($\times\frac{2}{5}$). **Right:** *Jay* ($\times\frac{2}{5}$).

The Tawny Owl is the only owl likely to reside in urban and suburban areas. For success it requires mature timber which will provide holes of sufficient size for nesting-chambers. Householders first become aware of the presence of tawny owls in a vicinity when, in the early part of the year and after dark, a tremulous 'hoo-hoo-hoo' of considerable carrying power issues from the depths of some ancient garden or roadside tree. The noise comes from a cock bird and will probably be answered by a sharp 'kewick' from its mate near by. In the day-time, a tawny owl, now silent, can sometimes be spotted roosting on a branch close to the trunk of a tree, its mottled brown plumage making it difficult to isolate against the bark, especially if this is partially overlaid with ivy.

In woodlands, tawny owls live mainly on small mammals along with a few birds and insects. Near to dwellings sparrows and starlings form a significant proportion of their prey.

Members of the crow family appear occasionally in gardens but are rarely regular breeders. The Carrion Crow is a scavenger. It haunts central squares and will nest where there are lofty elms, sycamores, or planes. From time to time it visits suburban gardens, first alighting on a tree to assess the possibility of obtaining a meal before making a brief sortie to the ground. Uniformly black in colour, its call is a distinctive hoarse 'kaaah'. In spring carrion crows are a menace to other breeding birds such as finches, since they take many eggs and young.

Within woodland, which is their natural habitat, Jays are shy and rarely leave cover. Their presence is usually inferred from the harsh 'shaaak-shaaak' calls. Where parkland exists in cities the birds can be common and, when this is so, they lose their shyness and become bold. Although acorns form a major part of their diet in winter and, like carrion crows, they ravage

Below: *Magpie* ($\times\frac{1}{4}$). **Right:** *Rook* ($\times\frac{1}{3}$).

the nests of smaller species in spring-time, jays frequently make marauding sallies on suburban gardens. Broad beans seem to be particularly attractive. Birds hop rather heavily along rows stripping beans, one at a time, from the pods.

Jays are the most brightly coloured of the crows, pinkish brown on the body with darker wings enlightened by the bright blue splash of the wing coverts.

The possibility of seeing Magpies in gardens varies in different parts of the country. They are still less numerous in the south than in some industrial parts of the north where they can be familiar visitors to suburban gardens for kitchen scraps. Like jays they are very wary in their natural habitats of hedgerows or thickets, but where garden-visiting has become established some of their basic caution disappears. Long tailed and pied, they cannot be mistaken for other birds and their chuckling 'chak-chak-chak' also allows them to be easily identified.

Rooks nest communally in rookeries which are often of great age. Few gardens accommodate them and, as with carrion crows, they are more

Above: *Pied Wagtail* ($\times\frac{2}{3}$).

likely to be occasional visitors looking for food. Although predominantly black in colour, rooks are differentiated from crows by a white area on the face. In addition the feathers round the tops of the legs are rather loose and give a baggy appearance.

The Pied Wagtail makes casual visits to many gardens in the outer suburbs, especially if they are adjacent to open farmland. It appears most frequently at the end of the nesting season when families are dispersing, a slender black and white bird whose tail is forever pulsating up and down. Usually only single birds are noted busily flitting and darting over lawns, snapping up insects, good numbers of which are available in late summer.

Where old-fashioned walled gardens still exist wagtails are much commoner. They build their nests in recesses and the space created by a missing brick is often taken over. Normally walled gardens are large enough for a mound of decaying manure or compost to be tucked away out of sight. Being ideal breeding-grounds for flies, wagtails find them good feeding-stations.